The Little Guide to Empathetic Technical Leadership

Alex Harms

The Little Guide to Empathetic Technical Leadership

Alex Harms

ISBN 978-0692674857

Published by Maitria

Making the world a little gentler for developers & tech teams. Geek joy, mindful code and mindful human connection.

The Little Guide to Empathetic Technical Leadership

Introduction

A leader is best when people barely know they exist.

When the work is done, the aim fulfilled,

they will say: we did it ourselves.

—Lao Tzu

leadership

When I think of leadership in relation to self-organized teams, other ideas come up as well. Power. Authority. Autonomy.

In fact, thinking of any organization or organizational structure will eventually lead me to thinking about those things. Who decides?

I have been told I have a problem with authority. What about you? When you think of authority, what comes to mind?

When you mull over "leadership", how does it feel different from "authority"? How are they the same? I often hear the words "leadership" and "authority" treated as synonyms. Let's untangle them a bit.

In an organization, someone with authority is empowered to make decisions, and those decisions are considered binding on other folks. Authority is *obeyed* or *disobeyed*. Authority comes with power. The power to hire, fire, administer bonuses or "disciplinary actions".

What happens when an empowered technical team decides to make decisions together? To what extent is it possible for them to keep power-based authority outside their day-to-day work? Is it a good idea?

And what's the role of leadership on such a team?

who decides?

As a student of humans, some of the most interesting lessons life has provided have involved my kids. People would tell me "why don't you make them {action}". I just kept thinking "How would I do that? How could I *make* someone do anything?" The answer became pretty clear. You make them, someone explained to me, by devising a punishment that is severe enough to outweigh whatever benefit they're getting from doing what they originally chose, so that they will choose to do what *I* prefer, instead.

As a young parent, I wasn't sure I wanted my kids to make their decisions about how to behave based on fear of my anger or punishment. More importantly, I recognized in them a decider-ness, a self-authority, that I didn't want to break.

What amazing things we humans are, with our ability to

consider a situation, mull over possible outcomes, and *decide* what we *choose* to do!

And in the end, it's clear that every person decides what they will do, anyway. I can try to set up the environment to steer them toward the choice I prefer, but they will still be constantly making choices. *Do we build the world we want while fighting that impulse, or embracing it?*

who decides what?

What I have for lunch, for example, is entirely up to me. Within constraints, I get to decide. Constraints include things like how much money I have, whether I cooked something earlier, whether I'm willing to steal, whether there are donuts within two meters of my desk, whether the restaurant I chose is open. But whatever the constraints, the decision is mine to make.

You can try to convince me to have something else, and then you're recognizing my ownership of the decision. I love when that happens, because I want to learn things, grow, make better decisions. I want you to try to convince me when you think you have a better idea!

Or, instead of trying to convince me, or offering me alternative ideas, you can physically force me to do something else, and then you've become one of those constraints I was talking about. The constraint of whether some jerk is physically preventing me from eating my lunch, or not.

And what you will have for lunch, similarly, belongs to you.

The organization owns the decision about who it's willing to work with. Who *I* will agree to work with is a decision that belongs to me. Constraints might be pretty intense, like whether I have other options for food money, or housing money. Still, the decision rests with me. That's something no employer has *authority* over.

When we get to topics like what time I show up, who decides?

Well, I can tell you this. I decide what to do with my body. I decide whether to do things to get to work by 8 most days. My employer decides whether to continue the work relationship given the information they have about me.

I'm interested in how my employer feels about this. I'm interested in whether my employer wants to continue the relationship. I might choose to come in earlier than 9:15 to help the employer be happy. I might also want to know what makes them happy about 8:00, and whether we can meet that need some other way.

Know what I'm not interested in? I'm not interested in cooperating with tactics intended to "alter my behavior". To me, the idea "being a human being" has in it the idea "working together to solve problems" and "caring about one another's wellbeing". It does not have in it "being molded to fit someone else's idea of who I should be".

organizational collaboration

When I first heard about Agile software development, I was excited and inspired. There are people who advocate self-organized teams, developing software using their whole, creative selves! Not doing what they're told, like automatons, but working together by discussing things, improving their working models, choosing, and then recursing until they return a working product! How lovely is that? Suppose we can get everybody involved to agree that we're all equals, and want to make our decisions together *by thinking things through* together, rather than by attempting to use power and coercion on one another. What does this sort of work, within an organization, look like?

What do we do with the concept of authority on such a team? Where does the power reside? *Who decides?*

We all want the company to do well, so we consider the company's interests. Often, those are represented by a manager, for example, but the manager has stuff they value individually, too. And then, of course, we consider the interests, or needs, of each individual, as well. And we listen to one another.

If I feel like I'm at choice, like I'm an agent, empowered to do what seems best to me in every moment, and the people around me also feel that way, I enjoy work. I'm engaged. Yay! Not only do I enjoy it more, but we make better decisions together than any of us could make alone. I'm pretty smart, but I would not like to see how my life — and my work projects — would be different if I didn't learn continually from other people and the ideas they bring me.

what is leadership, then?

So without power-based authority, what does it mean to be a leader?

How many people are there in your life who can say "trust me" and tell you to do something without explanation? How would you respond if the person you're closest to was eating in a restaurant with you and said "Don't turn around. Just stand up quietly and walk out the side door." What if a police officer said that? Your boss? Your grand-boss?

When a boss says "trust me", it might not be a request to trust. It might instead be another way to say "Don't question me." One way you can tell this might be happening is that when you comply, you feel your eyes roll a little. The stories you tell later make clear that you weren't inspired or persuaded.

Think of a time when your favorite leader from a TV show or a book said "Trust me." The character has been developed to indicate that trust is already there. The leader has earned some authority, or decision making influence. Sometimes this is called "charismatic authority", but that phrase ignores the agency of the folks who implicitly grant that authority. When you trust someone that way, you're *giving* them authority. So I call it inspired authority.

When authority comes from law or from tradition, it risks gaining *compliance* at great cost. While we're taking the "safe" route, keeping the peace, avoiding uncertainty, we lose information and opportunities for growth. We miss out on better outcomes.

In this Little Guide, when I talk about leadership, I'm talking about the ability to inspire. Sometimes this looks like inspired authority. Much more often it doesn't look like much at all.

Here are some clues that you might be a leader:

- people come to you with questions
- you provide more guidance and encouragement than answers
- you listen more than you talk
- when you talk, people listen
- when something worries you, it worries your team
- you hear from teammates when there's a problem
- they confess both their fear and excitement to you

A leader doesn't need imposed or structural authority. In fact, it may get in the way, if you're not careful.

what to expect

Technical teams exist to create something of value. We have a craft. As a leader, you study your craft, build on your skill, try new things, and share them with others. Sometimes those new things are horrendous disasters, but it's worth the risk, because sometimes they turn out to cut your work in half.

Every bit of software we write is new, even if it's similar to something we've done before. (Experienced developers knows what "just make it exactly like the old one" means. It means your customer has no idea what your job is like.) Learning is constant. And sometimes, learning is hard.

New tools, new testing strategies, new customer strategies! New methods, frameworks, even new languages and platforms. As a leader, you've more experience than many on your team. How will you share your experience? How will you continue to get better at learning, while you help your team get better at learning as well?

How will you manage all this technical work when other humans keep complicating things?

This Little Guide is divided into two sections.

being present

If only you could become an empathetic technical leader this way:

```
brew install empathetic-technical-leadership
```

But nope. To become an empathetic technical leader requires going to the hard place, developing what's inside you.

The first section is about the inner work you can do to help you be present for your team. Understanding emotions and building listening skills are the first steps to cultivating connection and trust.

working together

As a caring leader, you help to create the collaborative space for the team to thrive in. That space — physical, mental, and emotional — is the foundation that a team needs for learning to make decisions together, improve together, and deliver excellent software together.

Lots of communication needs to happen on an empowered technical team. Shared values, shared constraints, even a shared codebase means the communication never stops. And decisions need — at a minimum — everyone's real buy-in in order to work, to serve the team's ultimate purpose, delivering awesome products. Meetings happen, asynchronous communication happens, and conflicts happen. As a leader, you learn to facilitate well, to allow every voice to be heard, to avoid false consensus, and to help ensure each of those meetings or discussions improve team cohesion, rather than damaging it.

The second part of the book is about applying the skills in part one to your work as a technical team member and a technical leader.

In this Little Guide, you'll read about a leadership that's grounded in being fully present, with your whole heart. The foundational skills for this work are the hardest part. So we'll start there.

18

I

There's no reflection in moving water.
Only when you know internal peace
can you give it to others.
—Lao Tzu

The hard work of leadership starts with connecting with people. And *that* work starts with getting to know yourself.

listening

Some years ago, I was watching my kid's first classes in martial arts. About six months earlier, I'd stopped using a wheelchair for daily living. My disability was less severe, but still present. The teacher saw me fascinated by what I was watching in the introductory lessons, and even mimicking some of the movements, and came over after class to invite me to try a class as well.

Over my protests — "I wish I could! But I can't." — he explained that everyone decides for themselves what they can do, and is expected to take care of their bodies, and not push too far. Finally persuaded, I had great fun over the next few years studying at the dojo, and I learned, among other things, how to live with the contradiction between going farther than I believed I could, and going no further than my body could handle.

Of all the teachers, Sensei was the busiest person around. Our beginner classes were usually taught by another teacher, and only occasionally by the Sensei. But sometimes there was reason to talk with him, and on those occasions, I noticed something. When you were in a conversation with Sensei, it was like there was nobody else in the world.

That might sound like an exaggeration, but let me tell you. The kid in question was (and is) my most "self-directed" kid, the one who challenged everything, asked *why*. When there was something I thought it was important for him to do, he often had a better idea. :) This kid said, "There's something about Sensei. He makes me *want* to do what he's telling me".

I puzzled over this, but over time I got to know our teacher better. What I noticed was that when he called you into his office, talked to you in the aisle before class, or approached you on the mat to talk about the shape of your fist, all of his attention was right there, in that moment, in that conversation.

I'd experienced deep listening in special contexts, of course. An emotional conversation with a loved one. A weekend retreat where we're all about paying caring attention to one another. But this was different. This was every-day life. I thought maybe this was the secret to Sensei's power to influence people.

Afterward, watching leaders in other contexts, I saw a difference. Without that focused attention, I could see their effectiveness suffer. More chaos, more stress. I looked at leaders who inspired their teams, and each time, I saw the same attention, the same focused listening. Coincidence? I think not!

We all know that correlation is not causation. *Causellation* requires thinking. Understanding. So let's think about why listening might make the difference.

the experience of empathy

My first exposure to the idea of intentional listening was reading Carl Rogers in college. Rogers, known as the founder of Person Centered Psychology, called the people he worked with "clients", rather than "patients". Instead of diagnosing illness, he wanted to work with clients to help them become happier and more fulfilled — to *self-actualize*. His work has been extended to people in a variety of helping relationships, from teacher-student to coach-client, to manager-employee.

Rogers, in looking at what created the interactions that were most helpful to clients, most effective in helping them to grow, identified the conditions needed for "therapeutic change". These three have become known as the Core Conditions:

- The person in the leadership role (in his case, thera-pist) has **empathic understanding** of the "client".
- The leader is authentic, genuine. There's **congruence**; they aren't acting.
- The leader shows **unconditional positive regard** for the person they're working with, within the helping context. Fondness.

Rogers' experience resonated with me, but I didn't know what to do with it. (That third one, I thought about on and off for over a decade before it started to make some sense to me. I'll leave that as an exercise for the reader.)

Shortly after reading Carl Rogers, someone introduced me to the work of Marshall Rosenberg. Back then, empathy wasn't talked about nearly as much as it now, but Rosen-berg's work gave me a whole new way to think about it. From my work with his Nonviolent Communication[1] practice, I began to notice that when someone experiences

---

[1] Rosenberg was a student of Carl Rogers, who left psychology en-tirely, for similar reasons to Rogers' rejection of earlier approaches to the field of psychology. He created Nonviolent Communication, which you could consider an extension of Rogers' Person-Centered Approach, but for everyday life.

receiving empathy, feels truly heard, understood, and accepted, something shifts in how the conversation moves forward. The most amazing part to me was seeing that very often, once they feel heard, the person becomes better able to empathize with others, to practice self-empathy, and even to think things through and problem-solve.

It's a bold claim. I'll tell you what Rosenberg told me: please don't take my word for it. Try it in your own life, in your work, and see what happens. That applies to anything I tell you that looks like advice. Experiment with it. Find out what happens. Decide what works, what you like, based on your own experience.

the skill of listening

As you might guess, the kind of listening that creates strong empathy, that leads to connection and growth, isn't like everyday listening.

There are lots of different ways to listen. Way on one end of the continuum, you find half-listening. You're sitting with someone at lunch who won't stop talking. You nod, wishing it were over. Maybe feeling a little guilty for not paying attention.

Then there's listening while waiting for your turn to talk. Watch the next energetic discussion that happens nearby (or that you're involved in). See if you can identify this pattern. Closely related: listening until you learn something that interests you, then jumping in. Listening for the facts, and arguing when you disagree.

These kinds of listening happen without really trying. They're automatic. Now I want to talk about when listening is intentional.

focus

Suppose you're walking with a friend late a night, in a city. It's rained recently, but now the air is comfortable, and the streets glimmer with city lights. You're talking together about an interest you have in common, and then your friend stops, looks off in the distance, and whispers, "listen". What happens inside you?

Of course, the first thing is that you stop talking. But I'd suggest there's more. You also stop thinking about what you're going to say next. You start sensing the world around you. You're right there, in that moment.

"I don't hear anything."

"It's a clicking, or ticking? Coming from over there."

You listen, and suddenly you hear it. You can't un-hear it.

This is a part of what listening involves. Focused attention.

When we listen to another person, we are pretty good at picking up the surface meaning without really paying much attention. The conversation you were having with the friend, above, for example, was going fine before it was interrupted. (By a woodpecker, if you're wondering. Your friend had never knowingly heard one. "You mean those are real? I thought it was just a cartoon!")

Then you were called to focus your attention.

In a conversation with another person, an empathetic leader will often choose to listen with that sort of focus.

Focused listening involves some challenges. Here are two of them: harnessing your attention, and dealing with emotions that come up.

When I want to listen with focus, the first thing I need to do is to check in with myself about whether I can set aside my own concerns for the moment. I might be coding, writing, in the middle of a game, or worried about the health of a loved one. All sorts of things are on my mind, most of the time.

If someone says "can you talk?" I have a chance to consider whether I'm able to set those things aside for another time. Plenty of times, I ask to have the conversation later, when I can give my full attention. Sometimes, I'm able to set my stuff aside, and focus.

Even after I get past that hurdle, though, thoughts and emotions will come up for me in the midst of listening that are hard to ignore.

All through the conversation, I'm noticing when something comes up for me, and choosing to set it aside while I listen to what the person in front of me wants me to know. This is especially difficult when the things the person wants me to hear are about a conflict we are experiencing.

Let me say that more clearly. Sometimes, the person is expressing themselves by yelling, or they're venting about things that affect me as well. Sometimes things get emotional, or heated. Those times I'm able to stay present and listen, without my own emotions getting between us, I get connection and results that make it all worth it.

listening to understand

Focused listening will let you hear what the person is telling you. Listening to understand will help you— well, understand. Have you ever had a conversation that went roughly like this one?

"If you'd pick your socks up, we wouldn't have this problem."

"You're telling me it's about the socks?"

"I didn't say that!"

"Um, you just said that."

"Uuuuugh. Why do I bother?"

You've heard every word. But you haven't *heard* anything. What went wrong?

Here's an idea to chew on. When you're listening to another person, they have something they want to communicate. Ok, that's obvious, but stay with me.

What I'm saying is that they're excited about something, or worried, or sad or happy or... *Every* time someone tells you something, they're feeling something, and it's an impetus for them to speak to you. Seeing beyond the thing they're telling you right now, to what's underneath it, is sometimes a very helpful skill to have.

When you get to listen not only for understanding of the ideas, but for connection with the feelings being expressed, you may find real surprises. A conversation can start with "We really need to get Marcus off the team," and end with "Oh, yeah, I think I would like to pair with him more."

Don't believe me? That's cool. You know my mantra: try it for yourself, and see.

developing your skills

Earning trust, building connection, facilitation — every skill covered in this Little Guide starts with listening. Developing the ability to choose where to put my attention made such a tremendous difference in my ability to connect with people; it's also some of the hardest work I've done.

But, fear not!

In the next few chapters, we'll talk about how to develop the ability — within yourself — to listen as an empathetic leader.

mindfulness

Listening with empathy is hard work. This chapter is about getting prepared for that work.

Mindfulness practice is sometimes offered as a relaxation technique or a way to de-stress. It can accomplish those things as a side-effect, which is lovely. But for me, the more interesting part involves how mindfulness gives me insight into the workings of my mind, makes it easier to question my assumptions, and gives me more control over my attention.

Before I say more, let me invite you to a brief exercise in noticing. The exercise is a simple five-minute mindfulness practice. I'd suggest reading these next couple of pages very slowly.

When you're ready, you can begin by turning the page.

noticing

While you're looking at this text,
move your attention to your body.

Notice the sensations in your hands, your feet.
Notice the feeling of gravity holding you in place.

Do you notice any sounds around you?
Voices? Outdoor sounds? Ventilation?
Notice what you hear, and let it go.

Notice any smells you perceive. Any tastes.

Just notice, and let the awareness pass by.

Without trying to change it,
notice how your breath feels
moving into your body, and out.

Notice what your breathing is like.
Is it rapid? slow? shallow, or deep into your belly?

Just notice, without trying to fix it.

Let yourself fall into a rhythm of long, slow breaths.

Begin to notice whether thoughts are coming to you.
Are there thoughts of things that happened earlier?
Things that might happen?

Without dwelling on the thoughts,
notice them, and then let them go.

Bring your attention back to your breath,
and its long, slow rhythm.
Watch the breath. Notice the sensation of the breath.

Begin to notice your other bodily sensations.
Are your shoulders relaxed, or tense?
Is your jaw open, or is it closed?
How about your neck?
Where else do you sense things in your body?

Notice the sensations.

And turn your attention back to your breath.
After a few slow breaths,
start to notice any emotions coming up.

Are the emotions connected with sensations?
Let yourself linger on the question.

Notice if they're connected with thoughts.
Let the thought and the emotion pass, together.

And bring your attention back to your breath.

In this exercise, you had a chance to practice shifting your attention several times. As your attention moved from your body's sensations to your environment, from your environment to your breath, were there any times it was difficult? It may be easy during relaxed moments, and yet become nearly impossible during moments of stress, to turn your attention to something other than your mind's rehashing of the ideas that are upsetting you. Practice — lots of practice during times when it's not so rough — can help make that shifting of attention easier.

My suggestion to you is that the challenging aspects of listening, the part where you choose to put your attention on the person you're listening to despite your own pressing topics, and especially the part where you stay present even when something brings up emotions for you, are so much less difficult when you have a habit of practicing mindfulness, however you do it.

You know what comes next, right? Say it with me: Don't take my word for it. Try it, and see what happens.

The rest of this chapter is about mindfulness, emotions, and gently teaching yourself to respond from a place of caring, rather than reacting from a place of fear.

meditation & mindfulness

When you think of meditation, maybe Hindu or Buddhist meditation, or even western yoga, comes to mind immediately. But Christian, Jewish, Muslim, and other traditions also include some form (or forms!) of meditation. Styles vary; intended outcomes vary. The word meditation (like so many words) doesn't have one, clear definition.

Mindfulness is easier to think about. Jon Kabat-Zinn, an early popularizer of mindfulness practice in the West, called it 'moment to moment non-judgmental awareness.' Did you experience some of that during the exercise?

Some folks achieve that through hours, even years, of zen meditation. On a little black cushion, in a temple on a mountaintop, or in the corner of a bedroom. Others practice in a church or a mosque, or on a park bench or in a garden.

My journey started when I began to notice how difficult it was to slow my mind down, to observe more than just the thoughts swirling around in my head. I read a long list of books about meditation and mindfulness, but I kept *not doing it*. Finally, I decided to find a group of people who were doing it. I felt it might be easier to remember not to get up if I were surrounded by other folks not getting up.

I went to a zendo, and for a few years, my life was a combination of daily meditation, martial arts, and walking or riding through the park, paying attention. Well, that's not quite right. It also included cleaning up kid messes, arguing with my spouse, bouts of hopelessness, and moments of joy. But in between those things, lots of the other stuff.

Gradually, I began to get a sense of what the zen masters meant about mindfulness in each moment. I began to be less mystified by teachers like Shunryu Suzuki, author of Zen Mind, Beginner's Mind, who said "Treat every moment as your last. It is not preparation for something else."

There are a few ways to practice mindfulness on a regular basis, so that during moments of stress, the habits are there. One way to practice mindfulness is just to do that noticing exercise from the beginning of the chapter. Take a few breaths, and start to *notice*. Another is to blend that practice of intentional attention with daily activities. Quietly wash dishes, noticing how the water feels. Noticing where your hands are. Or take a walk, alone, and pay attention to each step, without focusing on the destination, or what you were doing when you left, what you will do when you return.

Or, you know, hit the cushion and sit, zazen (zen style). As the zen masters say, you should sit for twenty minutes a day, unless you're too busy. In that case, make it an hour.

emotions

I remember a time when, if someone asked how I was feeling, I would be at a loss for an answer. If I were very angry, very scared, or even very happy, I could recognize those things. But gentler emotions were a mystery. And without access to those emotions, I had two problems. First, I didn't understand my reactions to events — and people — around me. I didn't notice my defenses kicking in, didn't know why I wasn't the person I wanted to be. Nor how to change.

The second problem didn't become apparent until I'd begun to solve the first. The second was that I wasn't able to empathize with other people's emotions or needs, until I could listen to my own.

Reading a book about being aware of your emotions, of course, is like reading a book about playing tennis. If the words are to be at all useful, that will come with practice.

For me, the breakthrough that let me pay attention to all of my emotions came on a rainy night at a weekend retreat. The lessons that day had been "All of your feelings are okay," "Your feelings don't dictate your actions." I was sad, and scared that there was something wrong with me for feeling so awful. For not being useful, normal, functional. I didn't see how all my feelings could be okay.

A teacher invited me in, and I sat in front of her in tears, explaining that I'm not okay as I am. That my feelings are not okay.

She said "They're okay with me."

"I know that's the offical position. But I'm not talking about normal feelings. I mean I'm sitting here sobbing. I can't stop."

"Yes."

"What about... sometimes I get really mad."

She said "It's okay with me if you're enraged. Sometimes I feel very angry, too."

"Not this angry. I get *too* angry."

"Even if your rage is new to me, it's still okay with me. Your feeling angry is okay with me."

...

"What if I'm jealous? Needy."

"Those are feelings. They can't hurt me."

"They hurt people I love."

"Do they? How could something that happens inside you harm them?"

My mind, as they say, was blown. More time, more breathing, more *questioning*, but gradually, I adopted some models that have made listening and connection with other people possible.

Or maybe they're hypotheses? I can't test them directly, but I have tried them, noticed what it meant for my life. Found them to hold up.

Consider these ideas. Try them on. Do your own experiments. If you think they might make your life easier, practice looking at emotions through this lens, and find out what looks different.

- When I accept feelings that come up in me, let them be without opposing them, pretty soon things shift. I never stay stuck.

- My feelings are what's going on inside me. They have to do with my own past experience, who I am, and ideas I have. My feelings aren't a direct result of someone else's actions.
- My feelings have something to tell me, if I listen to them.

All of those are about my feelings. But as a model, it applies to when other folks are experiencing feelings, as well.

an important message about support

You may not have as much to work through as I have had. Or you may have more. One of my favorite teachers says that if you're going to study loving-kindness, you'd better get ready; you're going to get to know yourself really well.

It's important to note that emotional work can become very difficult. Please remember that there are folks around to help, if you need it. If you begin to suspect that the work you're doing with mindfulness and emotional awareness is more than you can handle on your own, please consult your trusted advisors, clergy, doctor, or a therapist. Listen to your loved ones if they become worried. Nothing here is meant to take the place of your support system, and medical care (including mental health care) when it's needed.

empathy

When we empathize, we're recognizing a familiar experience. When I see someone hungry, I can be sad for them (sympathy), but I can also recognize hunger as something I've experienced as well. I can empathize. I can notice that although, for example, I'm not fond of their strategy of stealing my sandwich, I can understand the hunger that led to it. Now, rather than wanting revenge, I have ideas! Maybe we could get a vending machine. Maybe when I learn more of the story, and empathize more, I'll want to bring two sandwiches from now on.

Empathy comes from a place of deep human connection. Sometimes it feels easy. Other times, it calls on all the skill and love and openness I can bring to the conversation.

Did you ever wonder why airlines tell you that in case of emergency, you must put on your own mask first? Even before your little baby?

If oxygen levels drop, before you die, you sleep. Before you sleep, you get clumsy and confused. So imagine. While you're trying to get the mask for your baby, you become clumsy and confused. It takes longer. Then you finally get the mask on baby, but by then, you are too sleepy to get your own mask on. Now baby has a mask, but no grownup to take care of them.

If you put your own mask on first, you can easily put a mask on your confused or sleepy baby.

You're going to be surrounded by emotionally clumsy, relatively unaware or stressed people. If your own mask is secure, you're in a much better position to help.

self-care

Sometimes, I find myself I'm reacting to multiple demands on my attention, all day long. I'm scrambling to keep everything going, taking care of work, juggling finances, worried about kids or parents.

Sometimes, I'm exhausted, but pushing through anyway. Or I'm sick, but it doesn't feel okay to take time to recuperate. Sometimes I'm really scared, about the health of someone I love, about not being able to make rent.

When I'm in a stressed state, I've come to notice something. My ability to listen, empathize, and care for other people is impoverished. My energy is depleted, and because of that I don't have energy to spare for focusing on other people — on my team.

I notice I tend to be short with people, emotionally fragile, reactive. I don't even remember that I intended to be empathetic.

What patterns do you notice in your own reactions, when you're living with stress?

If it's hard to answer, try watching yourself over the next few days. Notice when you aren't as present as you'd like to be. When you're short with someone, or unhelpful. What circumstances create stress in you? How do you find yourself reacting?

If I want to increase my chances of responding in the way I'd like to, I find self-care is vital. For me, this means getting enough rest, feeding my body in a way that nourishes me, learning to breathe in a free, unrestricted way.

It also means getting time outside, time alone, time to do things I really enjoy.

It means pulling myself out of a project I'd happily work on for 30 hours straight if I could, and doing that *before* I'm forced to by bodily functions. Pulling away in time to get and give hugs, to notice the sunlight, to hear about my lovie's big news.

What are the ways you find yourself neglecting self-care? What would you like to change, about how you take care of your own needs?

empathy as self-care

I said earlier that feelings can tell us something. I'll go farther, and say that I've come to count on my feelings to point to something beautiful that needs my attention. Here's what I mean.

If something's bothering me, I might find myself doing any number of things — eating ice cream, snapping at people, frantically cleaning the house. If I notice I'm acting strangely, I have a chance to stop avoiding, sit still for a minute and breathe. I can let the feeling rise up, welcome it. Find out what sort of feeling arises, and what it wants to tell me.

The journey may be pretty unpleasant. I stay with it, knowing that the feeling isn't going to harm me, that it's safe to just sit and feel. Even more, I know that avoiding it isn't going to make it go away.

I feel a sadness come up. Staying in awareness, I feel the sadness, and I begin to notice what it's tied to. Thoughts come up. I wonder, maybe I'm sad about what happened earlier? Maybe Mom? Maybe it's that my partner is leaving for a long trip tomorrow. Now I feel something in my throat, tears welling up in my eyes, and I know I've hit on something.

Remember those three hypotheses at the end of the chapter on mindfulness? One was when I let my feelings happen, and accept them, things shift. Despite my fears that my sadness would never end, it always has.

So, I sit and feel sad about my sweetie departing, for however long it lasts.

I said that the feeling would point me to something beautiful. Do you see it yet?

While I let the feeling happen, while my face is wet with tears, I ask myself what it is that I care about, that I value, that I'm missing, or fear missing. What's the beautiful thing behind this feeling?

It's the relationship I have with my beloved. It's making breakfast together, snuggling, talking about work. Those things are beautiful, and losing them has triggered the emotions.

Now that, I can see them, I notice in the edge of my awareness some light, some enjoyment of memories, some looking forward to new ones. There's the beauty.

The practice I've just described is what we call self-empathy.
In short, it's the practice of asking "what's that about?"
and discovering what needs or values are underlying an
emotion. Appreciating those beautiful things can help to
bring us back to feeling some peace.

Self-empathy is a skill that can be learned. And it can give
us the ability to care for our needs in a way that reduces our
reactivity, and increases our ability to respond from a place
of caring. Wielded expertly, it can provide the relief needed
during a difficult conversation, to allow us to keep listening
even when what we're hearing hurts. Not always — we're
not superheroes. Sometimes we need to take a break for a
while. But sometimes I surprise myself. Maybe you will, too.

listening for empathic connection

In the last chapter, we talked about focused listening, and
listening to understand and connect. Empathy is the bridge

that can get us from hearing to understanding, to reaching that connection.

Empathy for others is a lot like self-empathy. It involves listening, accepting, and looking for clues to the beautiful treasures that underlie the feelings.

Here's one more mind-blowing idea for you to consider, try on, see how it works for you.

What if every person you come across is doing the best they can to meet needs you can understand and recognize in yourself, even when their strategies are not to your liking?

What if we all have the same basic needs, deepest values, and desires? The desires for love, community, safety, creative outlet— there are plenty. What they have in common is that, as people, we all know the experience of those needs. They're the lens through which other people consistently *make sense*. Appreciating those things we all love is some-thing that ties us together.

And what if we're all, in our clumsy ways, reaching for those wonderful things as best we can figure out, given who we are, where we've been, what resources we have available?

If we adopt that view, what happens to our idea of "enemy"? Or "asshole"? When we shift to this perspective, what happens to our anger?

What happens to the way you talk to *yourself* when you try this model on? What happens to your ability to forgive yourself?

What happens to our ability to connect, human to human?

in practice

So what does real-life, in-the-moment empathy look like? A little story to illustrate:

Walking through the team space, Gabe overhears, "This is such bullshit." Noticing the expression of emotion, he checks in, pays attention. It looks like Eman is frustrated with the code he's working on. Gabe asks "Hey, would you like to pair on that?"

"Nah, I'm good."

Okay, Gabe thinks. Invitation declined. He moves on.

Later, he's getting coffee, chatting with Becky about some code they've been struggling with. Eman comes up.

"The whole thing is a total &$*%@ disaster."

Because he's had lots of practice, so it's become a habit, Gabe quickly makes some guesses, some hypotheses to work from.

Eman is frustrated, Gabe thinks. He'd like ease in working with the codebase. And maybe Eman is scared that the code is going to keep getting worse, that his teammates don't really care or don't know how to make it better. And maybe he doesn't feel safe talking about it, or like there's any point in talking about it. It's hopeless.

Gabe doesn't assume he's right in these guesses, but he knows that checking in by saying "Is it correct, good sir, to think that you are feeling frustrated, scared and hopeless right now?" is not likely to go well. Analyzing his feelings isn't what Eman is interested in. He's thinking about the $^*# code he's been fighting all morning. He's not interested in discussing his feelings, or what needs lie underneath them. Gabe's working theory, that Eman is having those feelings, and that they matter, will have to do for now.

Truth be told, Eman is kind of a pain. I mean, he's doing the best he can, but he's grumpy, and his interests are very different from Gabe's. He almost always refuses to pair. Gabe wouldn't choose to start up a friendship with Eman, outside work.

Sure, he wants to be friendly and supportive. (On a good day.) But mostly, he wants to be on a team that delivers high-quality software regularly, and he wants to enjoy it. Good relationships and communication within the team make that possible.

So what does support look like right then? Take some time to consider what you might do in that situation.

One of the benefits of an empathetic perspective is that it provides an alternative to other interpretations that might be tempting. Instead of thinking "Eman's being a jerk again", Gabe thinks "Eman would really like to have more ease in working with the code." (Gabe can appreciate the value on "ease", so he sees Eman as being similar, understandable.) So instead of becoming defensive, thinking "Hey, our code isn't that bad!" and opposing Eman, Gabe can focus on listening to what's important to Eman right then.

He listens to Eman a bit longer, letting himself become curious. He asks questions not to influence Eman, but to understand better. Then, he says "Would you be willing to pair with me this afternoon, and show me? I want to understand what you're seeing."

Eman might say yes, or he might say no. If he says no, there are other ways to talk about it.

Gabe has created in himself — and in the teamspace — an opening to learn about something that might help the team move forward.

leading with empathy

Empathy can transform conflict into connection. It's not a magic formula, and it doesn't work by fixing the other person. It works by changing what's happening inside you, opening you to new ways of seeing, new possibilities. (If it was easy, you wouldn't be a leader. You'd just be standing there, doing the easy thing.)

Leadership that comes from a place of empathy has the power to create connection — and from that, collaboration — that will astound you.

56

II

One who is too insistent on their own views
find few to agree with them. —*Lao Tzu*

What makes a group of people working together a *team*? How do they learn to trust one another? How do they solve problems?

gatherings

what are we doing in meetings?

As a team, we get together for lots of purposes, including:

- to understand our work better
- to make predictions about what will be done, when
- to solve puzzles
- to learn & grow

Some of these are planned, and some are ad hoc. All of them are things I want to do with my team. I like gathering to do those things, in ways that work for us.

What? I *like* meetings? Well, sometimes.

I also get "invited" to meetings I don't enjoy so much, and hope to avoid. Meetings that are focused on:

- extracting (or inferring) promises (we probably can't keep)

- "collaboration" on things I have no influence over
- "collaboration" on things that make no sense to me, or seem like a Very Bad Idea
- planning the un-plannable

These kinds of meetings may be scheduled by someone outside the team. Some of what you learn here may be helpful for organizational meetings, but we aren't going to focus on that. In this Little Guide, we're about working together as a team.

I'll mention that there's one more kind of meeting I'm occasionally expected to attend. Those are the meetings that feel like they are actually working directly against my values. Those are meetings where I sense somebody is trying to:

- make me like something I don't actually like
- lay blame
- create a "sense of urgency"
- threaten the team

If those are persistent— well, I'll tell you what someone told me: if you can't change your employer, sometimes you gotta change your employer.

the team's own gatherings

It's that first set of gatherings I want to talk about here. The meetings where we can get things done.

So far, we've put a lot of attention to communication between two people. But what happens when the communication needs to happen among 3 or 5 or 12 people? If we look at the team as a system, a whole, how do we help that whole be a better listener? A more effective collaborator? A productive contributor?

When it's a team I care about, and a product I care about, I have pretty high standards for our gatherings. I want them to be valuable. I want to *enjoy* them. I want it to feel like accomplishing something.

To that end, I want to help ensure that everybody is contributing as well as they can.

humans need

You know what I love about little kids? They haven't learned ways to squelch or hide their needs.

When a little kid wants attention, and my mind is on something else, I often just give them the tail end of my attention. My peripheral vision. As much as I love the kid, and want them to have what they need, I don't feel like being interrupted. I give them half-attention, brush them off. And what do they do? They just keep coming back, over and over. I might as well just get up from the computer and admire their creation in the first place. In the end, they'll win. They always do.

This is because kids *need* attention. They need it like they need food and water. And they don't yet have the ability to hide those needs from us to be polite and helpful. They gradually learn this, which is probably a good thing for survival. But when they're tiny, they don't have it.

Your team isn't three-year-olds, but they *are* human, and adult humans also have human needs. We have coping strategies for postponing those needs too, so we can tolerate pretty awful, un-useful meetings. Sometimes we postpone mindfully, because we want to focus on something. But sometimes we postpone by telling ourselves we don't really have needs. Then stress builds up, and *we expend energy coping that we could be devoting to creation.*

A well organized, well-facilitated meeting leaves room for paying attention to everybody's needs, including the very human desire to contribute, to be heard, and to create something meaningful.

positive, productive meetings

When I find myself saying "I'm glad I went", what sort of meeting was it? Thinking about the characteristics of an energized meeting, I notice I'm happiest when:

- I know why we're gathered.
- I am feeling at choice, and not "trapped."
- I care what happens as a result of the meeting.
- I have something to add to the discussion.
- I believe we're empowered to carry out our decisions.

How about you? What do you notice about meetings you like? What about the meetings you hate? Take some time to notice your responses. In the next meeting you attend, take some notes about things you observe. What would you like to do differently?

Once the meeting has started, there are more factors to consider.

- Will we finish on time?

- Will we get anything done?
- Will I get to say what I want to say?
- Will the lead just push through their agenda, or will we talk about it?
- Will everybody just go along, or will they provide good challenges?

We'll get into what to do about those things soon.

deciding

Have you ever sat through a technical discussion — the big kind — where you couldn't really tell what was happening?

I remember a meeting I observed some years ago that had me puzzled.

Someone had an idea that they wanted the team to consider — I think it was a way to reduce the complexity of the codebase, to improve the team's ability to deliver. And somewhere, the discussion got weird.

Here are a few of the things I witnessed;

- Questions were asked, and never answered.

- Comments were made that I didn't understand, and I didn't think other folks did either, but no one asked for clarification.

- Issues were brought up and responded to, but the issue kept coming up over and over.

- At the end of the meeting, folks had things to do ("research this", "talk to them", etc.) but there was no plan for getting back together or using the information gathered.

- There wasn't much discussion of how the meeting would go. The manager led, and seemed to make ad-hoc decisions.

Sometimes someone would make a comment, and I couldn't tell if it was an objection, a concern, or support for the proposal. Sometimes an issue would seem to be resolved, then brought up again anyway. Or was it a different issue? I came away at the end with the sense that the discussion was over, but I wasn't sure why, how it came to be over, or what would happen next.

Experiences like this one helped to convince me that meeting facilitation is one of the most important skills for a leader to have. How we decide together has a huge impact on how we work together. Before we talk about the role of the facilitator, I want to talk about participation.

reaching consensus

In a consensus, everybody agrees on an outcome that's best for the group. This doesn't mean everybody gets their favorite option, but everybody is heard, and everybody accepts, or consents to, the result.

The reason I care about consensus, is because without it, we face several risks.

- If the group is happy to let a leader — or even a majority vote — decide, we don't explore as deeply, so we don't learn as much. With consensus, we all share ownership. Consensus can help us reach better decisions.

- When folks don't feel a sense of ownership over the decision, they may be half-hearted in their work towards agreed ends. And if they're unhappy with the decision, they may disregard it entirely, or even sabotage the work.

- When folks don't feel heard, they lose trust in one another. The folks who feel their opinions weren't valued lose trust in the team. The team loses trust in their decision-making process.

Have you ever experienced your team setting "team agreements", and then lamenting that they aren't enforced? It's a trick question. When team agreements don't get carried out, I often discover later that there wasn't as much agreement as we'd been assuming. Getting stronger agreement in the beginning means enforcement doesn't need to happen. Instead, it's just a matter of paying attention, and adjusting as we learn.

testing consensus

When I want to check for consensus, I first state the proposal clearly. I've tried saying "Do we all agree on that?" and regretted it. Later, we didn't know what we'd agreed on. Now, instead of "that", I try to get specific.

Then, I wait for an answer. I used to take vague nods as an answer. Then someone told me straight-up, when I said "but we agreed!" that they hadn't agreed, that I hadn't actually got an answer from them. Now I sometimes find myself saying "no, this is important. I really want to hear a response from each person."

One way to check for consensus is thumbs up for yes, sideways for for acceptance with reservations, and thumbs down for "I'm really not okay with this as it stands." If you're seeking consensus, you don't move forward until that thumbs down has been resolved.

Another test for consensus is Fist to Five, also called Fist of Five. (See the end of the chapter.)

Checking for consensus isn't a vote, and needn't wait until the end of the planned discussion. Sometimes you might notice that the lively discussion you're seeing is actually what a friend calls "vehement agreement". Whoever is acting as facilitator (or you, if no one has been selected) might ask at any point whether there's a consensus, and test for it. The tests are to discover where we are, not to rush an end to the conversation. On the other hand, there's no point in going on if everybody's already happy.

courageous curiosity

Making decisions alone may be hard or easy. But it's always easier for me alone than when I'm making decisions with other people. Whether as part of a pair or a whole team, when I'm not acting as facilitator, I want to participate fully in group decisions. To do this, I need

- focused attention
- the ability to listen to the other folks involved
- a willingness to expose my ignorance, so I can learn
- and I need to trust my team or my pair.

I have the ability to influence each of these (some more than others). Taking care of myself and my needs beforehand means that when it's time, I can bring my attention to the problem we're trying to solve, and I can listen well. I can more easily remember that I value each person's voice (including mine, but not only mine). And I can choose courage, expose my ignorance, and trust the team.

But it's not just my state of mind that influences our outcomes.

In a discussion where there's strong disagreement, for example about a library or methodology decision, there are sometimes background issues affecting the discussion. Ongoing pressure from managers to "go faster", unresolved communication bottlenecks with designers or testers, or just devs feeling that what they're building is pointless. These things can all affect the decision-making process.

Courageous curiosity is something I can choose no matter what other folks are doing. It helps me to get through those barriers, and open up a conversation that's shutting down. Maybe it will help you, too.

If you feel yourself resisting, being irritated or closed down, try looking for some real curiosity. "What does she love about this testing framework?" "Why is he so strongly opposed to dependency injection?"

One of the quickest ways I've found to diffuse a tense situation is to stop talking and listen, for real. Follow that curiosity you've found. When folks are really heard and understood, that can be the key to helping them listen as well.

Next, we'll take a deeper look at some skills that help with the role of facilitator.

Fist of Five

When the facilitator wants to know whether the group is in agreement they ask, "Do we have a consensus? The proposal is…" And each person holds up from one to five fingers, or a fist.

Your team may come up with their own definitions. My starting point is this:

Five fingers represents enthusiastic agreement.

Four is agreement, with reservations.

Three fingers means that the team member is consenting, with reservations. Often there's a little more discussion, depending on how the team is feeling.

Two fingers (or one) means the person has serious reservations. More discussion is vital.

And a fist indicates a block. The message is "I can't in good conscious go along with this." It may mean "If this happens, I'm going to be looking for another job."

facilitation

When I hear people talk about facilitation and meetings, there always seems to be this idea that our goal is to get in, get something done, and get out. That we should leave feelings out of it.

Obviously, delivering is what we're here for. But we deliver well when we're working well together, and that's the bit that gets overlooked. Leaving feelings out hasn't worked for me in the long term.

What I've observed, time after time, on real teams, is that work gatherings are where a bunch of people come together and become a team— or they don't. Meetings are the hearth of the team. We get to know each other, let our guard down, learn one another's strengths and weaknesses, and build trust. As an empathetic leader, you may discover that facilitated gatherings are important tools for building team cohesion, and improving how you work together. And you may find that meetings gradually improve as everybody learns. That over time, trust is increased and team decisions become easier.

If you're curious about this idea that meetings can be enriching, try it out and see. I'd love to hear your experiences.

the facilitator's job

To me, the role of facilitator is precious. When we facilitate, we get to hold space for a group of people who care about something, who have feelings and needs and brilliant ideas and stuck places, so they can think well together. When it goes well, I know I've helped them come together, form a connection, and create something wonderful — or at least useful — together.

Facilitation, at its root, refers to making something easier. It's a core role of a technical leader, and a role that can be overlooked, or over-simplified. A skilled facilitator can make a meeting feel — and be — productive.

Here are a few of the basic things I do, as a facilitator, to help a meeting flow:

- Prepare (and modify) the agenda.
- Set up the room with awareness about how the setting can influence participation.
- Remind participants of the topic and the goal, when they stray.

- Determine how decision will happen, if there's not already an agreement in place.
- Set aside my own agenda; resist the temptation to participate in the discussion.
- Keep track of time.
- Make sure records are kept.

I've been able to get through some easy meetings with just these tasks in mind. Occasionally, in the easiest ones, I've even managed to participate in the decisions, without things falling apart.

But there have been plenty of meetings over the years where this wasn't enough. We'd lose track of the topic, or we'd make decisions I thought we'd agreed on, only to find out some folks were very unhappy, once it was over.

facilitating with empathy

As facilitator, I am in a unique position. While I'm interested in making sure the goals of the meeting are met (or adjusting them until they're goals my group wants to and is able to meet), I'm not attached to which way things end up. Specific outcomes aren't my concern, in this context. What decisions are made, what actions are planned, take a distant back seat to my strong interest in helping the team work together, here and now, to *accomplish*.

This frees me up to pay attention to the needs of the team, and the needs of its members.

I've heard folks tell me that we don't need to bring more empathy to business. That meetings work just fine without all that touchy-feely stuff. I've also seen people go into the bathroom and cry, seen decisions come out of meetings that most of the participants are unhappy with, seen teams reach agreements that are then forgotten. I want something different. Lack of communication and unspoken frustrations make a big difference in how productive I can be, and how much I enjoy work. I suspect that's true for others, as well.

I've discovered some more nuanced factors that, when I take them into consideration when facilitating, create much better outcomes. Sessions are more productive, and people are happier. Often, I need to muster all the listening and empathy skills I can, in order to pull it off.

Here are some practices for being a hardcore empathetic facilitator. Try them on, if you like, and see how they fit. I'd love to hear about your experiences.

skills of the hardcore empathetic facilitator

ensure a balance, that every voice is heard

I had a boss once who explained to a group of us that at *our* company, we only hire people who can hold their own. Everybody at *our* company knows how to speak up. And further, that people who can't defend their ideas don't deserve to be heard.

My experience is different. I've often heard some amazing insights come from the quietest folks, in the silences between big ideas and rebuttals. In fact, I've seen no evidence of any correlation between someone's apparent introverted or extroverted tendencies and the quality of their ideas. In any group, you'll have quiet folks and some folks who have a really easy time speaking up. You have people who are more or less articulate.

As the person in the room who has the big picture in mind, you can see what's happening from a broader perspective. There are some things you can to to maximize good ideas and the team's ability to hear them.

- **Set things up so that everyone speaks**, at least a little, in the first few minutes. Once someone's spoken, they're less likely to stay silent later. No need for subtle manipulation, either. It works *even if you tell the team why you're doing it*. "So, I'd appreciate it if everybody could just say hi, and tell us your favorite kind of ice cream." As the team matures, you can begin using more intimate ice-breakers. "...something we don't know about you", or even "...in one or two words, how you're feeling right now." (If you use that one, try putting a list of 4-6 feelings up on a whiteboard, to help them remember what "feelings" are. More on this later.)

- **Set the intention ahead of time** of balancing voices. Ask for folks to leave a little space between comments, rather than responding right away. Ask their *help* in making space for all voices. If the team knows your intention, they can help you out by not jumping in right away.

- **Watch for repetition**. If someone keep saying the same thing, it's often an indication that they aren't feeling heard. If you stop, look directly at them, and tell them what you understand their position to be, that will give them a chance to clarify, and they will have no need to continue trying to get their idea across. People who dominate the conversation often don't realize it, and don't realize that they've been heard.

- **Don't pressure quiet folks** to speak. You want them to have lots of opportunity, but pressure doesn't help ideas flow. Remember you're in it for the long haul. This one meeting isn't the end of all. You'll notice communication within the team will improve over time — often in just a few weeks.
- **Fully include folks who don't think out loud** by gathering ideas on sticky notes, then share.

Being explicit can be hard at first. Here are a few examples of things you can say directly, to get the team's help:

- "I'm going to be working on making sure there's a balance of voices. That means, roughly, everybody talks about the same amount."
- "Dan, before we hear more from you, I'd like to leave a little space for some other folks to talk."
- "Sandy, you look a little puzzled. What are you think-ing?"

Easy does it. Folk who are feeling introverted might have any number of reasons for not talking. Being explicit increases choice, and can help your team come to better decisions by getting more information, more perspectives. But you can't *make* somebody participate. Pressure will only increase their chances of shutting down.

It's helpful to pay special attention to status differences. Someone with perceived higher status will usually spend way more than their proportion of time talking during a meeting. Not only that, but their ideas will be accepted more readily. Think of the friction that makes it hard to start pushing a car. It's easier once it's in motion. Higher status folks are already in motion. Their ideas have an advantage. It's on you, as facilitator, to keep things flowing, and help the team come up with more ideas than one lead or manager can produce.

Lastly, it can help to understand the balance of the conversation if you read up on implicit bias and team communication. You may be surprised to discover there are systemic influences on how teams communicate. It's been pretty well established, for example, that women are interrupted by men far more often than men are. Most interesting to me was discovering that research overwhelmingly tells us that in groups, men talk more than women, and women are perceived as talking much more than they do. And it seems that when one-third of a meeting is women, folks estimate that numbers are equal. When half are women, people report that the meeting was mostly women. (I'll leave the googling as an exercise for you. But if you like academic studies, check out Interruptions in Group Discussions[2].)

As you read about it, you can watch your own meetings, critically, and see what you observe.

what problem are we trying to solve?

As facilitator, tracking the current topic is one of your primary jobs. Sometimes you'll realize during the discussion that the group has become distracted. It makes sense — everybody in the room comes in with their own needs and concerns, and new ones can arise during the discussion. Unless someone's attending to the structure of the conversation, it can become a cacophony for all those needs wanting to be addressed.

Any time you ask yourself "wait, what problem are we

[2]Interruptions in Group Discussions: The Effects of Gender and Group Composition by Lynn Smith-Lovin, Cornell, and Charles Brody, Tulane.

actually trying to solve here?" you can share that question with the group to help regain focus.

holding and modeling the group's values

I used to think that talking with tech folks about things like caring for one another would get me run out of town. Instead, I've discovered, over and over, that folks want to hear about it. (Not everybody, but more than you'd think!) Get this: once, at a conference, in front of an audience of 100+, I asked "Who here would like to be more loving, at work?" Probably half of the hands went up. Remember that lots of folks wouldn't raise their hand if you asked "Who here is a human being?" So half is saying something.

I was shocked, and it gave me a boost of courage in the work I do. I can't tell you what will happen if you try something that outrageous. Try it! Find out! Experiment, and let me know what you learn.

Your group may have explicit values, or implicit values. If they're explicit, you can pretty easily say "I remember the group deciding you wanted to do a lot less interrupting in technical discussions."

But even if they're not explicit, you can make them so. "I've had the feeling this team cares a lot about learning (...about test driving everything, or about listening to each other). Is that true?" (nods) "How does that affect how we move forward right now?"

I imagine if you state values you don't believe the team holds, it will feel like manipulation, and work against you. On the other hand, you don't need *proof*. Have a little faith. Can you see the team as caring about learning from mistakes, without blaming? Then, when things are getting hard, remind them. They may thank you.

Notice what's not being said

Have folks been telling you things in the hallway that don't match what you're hearing in the meeting? Are there issues you know need to be resolved?

Whether you're facilitating or participating, you can certainly speak about your confusion, about how things aren't matching up. And then, when you ask the question about what's happening, or whether the thing has been resolved — whatever the question is — *wait for the answer*. Learn to love that silence. Resist the temptation to fill awkward silences, and you'll learn a lot.

Disagreements are super valuable in a technical or other team discussion, because that's where the learning is. If everyone agreed with every proposal or idea, imagine what that would produce! As you gradually build trust, you'll develop into a team that's hungry for disagreement. Folks will say "I have this idea; please help me see where it's weak, so we can find a better version!"

observation

When I'm facilitating, one of the most useful things I do is to observe what's happening in the room.

body language

I learn a lot from the body language of the participants.

- yawning or drooping
- zoning out
- scowling or looking puzzled
- texting or emailing
- side conversations
- interrupting
- standing or leaning in

Listening and empathy give me clues to what's happening. Asking questions. Waiting for answers.

unanswered questions

As the group is working together, when questions come up, what happens? If you notice questions going unanswered, it can be a clue that the meeting isn't working well. Folks talking past each other could mean that they're confused, that they're upset and it's keeping them from hearing each other, or that they're feeling powerless. This is where we facilitators can help.

And, in case I haven't said it enough yet, when you ask a question, let there be silence for a bit, so that answers can start to come.

when things get weird

When things are stressful, the discussion can start to un-ravel. Sometimes — I'll tell you now — you won't be able to hold it together. A meeting will just feel like a disaster. You can't control all the factors. You can't make every meeting work; so I hope you won't kick yourself. Learn from it, and try again. But here are some of the ways you might be able to increase your chances of getting through an extra difficult meeting, when things get weird.

facilitation agreement

Sometimes, I've asked the team to enter an agreement with me at the beginning of a meeting. I ask what folks want from a facilitator, and they usually start with "end on time" and "stay on topic". Often we get to "resolve the issue" (not in my control, but I can try to help), and I'll ask them "Do we also want to make sure everybody gets heard?" Usually, they say yes.

Then I ask them for some things in exchange. They agree that I decide when we're out of time or need to move on, and they agree to respect the boundaries I set.

This is more structure than I often use, but when it's needed, it's been useful.

I remember one meeting vividly where I stopped and looked someone in the eyes (after the fourth refusal to relinquish the floor) and told him that either I would move us forward, or I would leave the meeting. It was scary, and I still don't know if it was the best thing I could have done. But it was the best I could come up with right then. The next day, he said two things to me: "thank you", and "I'm sorry".

My priority is on helping the team do what they need to do to deliver, and to grow as a team. Sometimes, it's not warm and fuzzy.

ground rules

Taking a few minutes to help the team create ground rules for their meeting can be really helpful. The real benefit is that it lets them make their values explicit. I usually start this process by asking "What do you want from the meeting?" and "How do you want to be treated?"

Just having those values in writing can help a lot in setting the tone.

nobody panic

If we take seriously the role of meetings in the growth of the team, it's important to remember that it's a long-term process. Don't panic.

Give it time, and watch what happens over a few weeks, or a few months.

retrospectives

looking back to move forward

Retrospectives are gatherings where we consider how we've been working, and look for ways to improve it. One of the principles behind the Agile Manifesto says "At regular intervals, the team reflects on how to become more effective, then tunes and adjusts its behavior accordingly." But the practice of looking back to find ways to improve didn't originate with Agile. The US military has a formal process called the After Action Review. And around 170 CE, Marcus Aurelius wrote this note to himself[3] on the subject of learning and changing:

> "If someone can prove me wrong and show me
> my mistake in any thought or action, I shall
> gladly change. I seek the truth, which never harmed

---

[3] The untitled work has come to be called, in English, *Meditations*. It seems to have been a private journal not intended for publication. Several versions are available free online.

anyone: the harm is to persist in one's own self-deception and ignorance."

This is the value — courageously welcoming information that can help us — that allows us to improve.

what could possibly go wrong?

We get a team of 6 or 8 or 12 people together to consider events of the past iteration (or whatever period they agree to retrospect about) and ask them to look at what could go better. What happens next?

What I've seen pretty often is that a team is told to "have retrospectives" — or they get the idea from reading or whatever — with no idea how to make those retros work. There's no other guidance than "get together to decide what to change". In the absence of specific study about retros, I typically see these patterns develop.

Retrospectives become blame sessions

Shame and blame can produce change. In my experience, those methods are slow and cumbersome. Continual, agile improvement comes much more readily from curiosity and even fun than from a sense of wrongness, or scrambling to avoid feeling broken. We have all sorts of strategies for avoiding pain that don't actually lead to improvement. And shame blocks creativity.

Fear prevents addressing real problems

When I'm introduced to a team that's experiencing a lot of fear, I'll often see a room full of elephants that apparently no one else can see. When I point them out before working to create a safer environment, it results in discomfort, rather than lively discussion. On the other hand, if we work on the culture or environment first, folks will begin pointing out those elephants on their own.

Someone pushes through their own agenda

Another pattern I see with casual retrospectives is that someone with power — a manager, or even a team member who is confident, loud, opinionated, maybe even trusted to make the technical decisions — will push to control the discussion and get their preferred outcome.

The first problem with this pattern is that their idea may not be the best outcome the team could have come up with. But the deeper problem is that the team doesn't get to *learn to improve together* if someone takes charge like that.

robust retrospectives

A *robust retrospective* addresses these problems and others by being both data-driven (so we learn from the world, to the extent we're able, rather than the ideas we have about it when we walk into the room) and empathetic (so that together we can be more curious than self-protective and scared).

data-driven

There is a popular (or at least common) style of retrospective, where each team member shares things that they think went well, things that went badly, and things they'd like to change, and then the group decides on some actions to try. These sessions can help a little, for a while. But as folks find themselves applying simple solutions to a simplistic view of the problem, or scrambling to fix symptoms without identifying the underlying causes, retrospectives start to seem less valuable. After several attempts to stop the boat from sinking by bailing, the team eventually decides to have fewer retros, or to have them only "as needed".

Without data, we can only guess at what is holding the team back. On a team that's trying to learn new approaches to software — like pairing, breaking down work into small pieces, or limiting work in progress — those guesses can be pretty far off-base. Data, including what's changed over time, can help us find patterns and look for likely causal factors that can be adjusted. Then we can form hypotheses and try things to see how our work changes.

A data-driven retrospective doesn't have to be heartless. Suppose you want a retrospective on the topic of feelings. In that case, you gather data about feelings, and start your team's exploration there. Try an app for tracking feelings. You may get surprised.

empathetic

Besides being data-driven, to be most effective, a retrospective needs to be a place where the humans involved feel safe enough to be honest.

Earlier, I suggested a model for understanding other people.

What if every person you come across is doing
the best they can to meet needs you can understand
and recognize in yourself, even when their
strategies are not to your liking?

Closely related to this idea that we're all trying to meet the same human needs as best we can, is the tradition that's emerged of starting a retro with a reading of the Retrospective Prime Directive, from *Project Retrospectives* by Norm Kerth.

> Regardless of what we discover, we understand and truly believe that everyone did the best job they could, given what they knew at the time, their skills and abilities, the resources available, and the situation at hand.

This is something a facilitator, coach, or team member will share with a team at the beginning of a retro, in the hopes of kicking off a shame-free, blame-free conversation about how the team can improve. Sometimes folks are like "ok, I can dig it" and other times there's a bit of an argument, or just silence. Because anger is a real thing, and denying it doesn't make it go away.

Building listening and empathy skills can help the team to avoid finger-pointing, and as a result, they'll conduct way more productive retrospectives.

We need retrospectives that are both data-driven and empathetic. Here's a structure you can use to plan your team's next *robust retrospective*[4].

structure of a robust retrospective

Planning a retro ahead of time will help you create an effective retro that digs up useful information, and uncovers valuable solutions to try.

When I'm starting to plan a retrospective, I start with this template, to help me remember all 6 stages.

1. Prepare
2. Gather the team
3. Gather or present the data
4. Examine the data and think it through
5. Decide on actions to take
6. Close

---

[4]This structure is descended from the excellent work by Esther Derby and Diana Larsen, *Agile Retrospectives*. I strongly recommend their book for a foundation in retrospectives.

1. Prepare

Choose a theme for the retrospective.

My retros have gone best when I've planned them ahead of time. Taking some time to sit and contemplate the team and its current state helps me know where to invite the team to put its attention. I let ideas bubble to the surface. What sorts of things could the team gain from thinking more about?

Before the meeting, the facilitator has a chance to decide on a theme for the retrospective. Are there concerns that you know, or suspect, need to be addressed? Are things pretty slow? Has it been a while since the team appreciated one another? Your insight into the team and the project will help you determine what to focus on.

If I've noticed a lot of uses of the word "velocity" recently, I might ask what needs lie under the concerns. Are folks concerned about whether the team is delivering as quickly as they might be? In that case, I might want to structure our retro as an exploration of that question.

Choose the data we'll look at.

For a meeting about whether the team could be delivering faster, I may want to gather data about how long it takes to do each bit of work. I might use cycle time[5], or number of stories done per iteration. I'd probably want to track this over time, so we could look for patterns.

If my retro were specifically about how folks are feeling, I might plan ahead to look for ways to track those feelings over time. We might want to look for thing that correlate with feelings as well. (Instead of tracking over time, we might gather data about those feelings during the meeting, instead. This could be less reliable, as feelings change often, for many reasons.)

Prepare the space.

Finally, just before the meeting, I decide how I want to set up the room for the session. How the space is arranged can communicate something about our intentions for the session. You may find, as I have, that you get much more engagement if you can set up in a circle, with no table in the

---

[5]Cycle time is a way to measure how long it takes to complete a unit of work. Though it's outside the scope of this Little Guide, there's a lot of information about this on the web.

middle. Tables can create distance, and encourage folks to be emotionally guarded.

For teams with remote members, setup is even more important. Retrospectives are one place the team gets to know each other, and that's extra important with remote members.

How can you reduce distance between people? If everyone's remote, you can make sure you're using an app that lets everyone see and hear each other clearly. What about when some are remote and some are on site? Have you experienced those meetings where there is a remote person on a big TV at the end of the room, and everyone else is around a table or in a circle? In my experience, this setup has resulted in difficulties in communication, and the person or people who are remote become distanced from the team.

The best solution I've found to this is to have everyone be remote, if anyone is. This means that even if you're in the office, you dial in just like the remote folks do. If you have trouble getting the team's buy in for this, maybe you can convince them to take turns working from home, so they can see what it's really like to be the remote person in a conference room. It's good empathy practice, and an opportunity to remind them how important it is that each member of the team know they're valued.

2. Gather the team.

Once we're ready to begin, my goal in the first few minutes is to help everyone bring their attention to what we're about to undertake, and set aside any distractions from outside. I also want to remind us of our shared values and care for one another.

Then I'll open with a quick ice-breaker exercise that helps to get everyone speaking, at least a little. Ensuring everyone speaks in the first few minutes helps to balance voices later on. When I've forgotten this step, the quietest folks sometimes don't say anything for the entire meeting. When I remember, I see more even participation.

As part of this stage, I'll also share the agenda with the team, make sure we're all in agreement about the structure and that the team wants me to facilitate. I don't much enjoy "helping" when the help isn't wanted!

3. Gather or present the data.

If I have gathered data in advance, here's where I'll share it with the team. If not, we gather the data as a team, right here and now. "Data" means the information we'll look at for insights into what we might want to improve.

One of my favorite ways to gather data when I don't know what the theme of the retro will be is to ask for help creating a timeline of the period we're covering. Participants write on stickies things that happened during the retrospective, and then place them at the point on the timeline where the thing occurred. You'll find out what stands out to team members, and then you can use that data as you move to the rest of the retrospective.

For all sorts of data gathering, I've found it helpful to have built skill among the team members in distinguishing our observational descriptions of a thing from our interpretations or evaluations of that same topic. When data gets mixed up with feelings, we don't have the distance to think clearly about what we're seeing. Feelings need to come up, but separately from the data gathering.

I've often been surprised at how difficult team members find this, especially when there are strong emotions in the room. Sometimes we get closer, but the judgment is still present. Then I remind myself that we're in it for the long haul. Over time, skill increases.

And as skill increases, retrospectives become more effective.

Observations and Interpretations

Data is true or false, not opinion. You can often record data with a camera or microphone. Feelings and judgments are the subjective part.

Separating data from the feelings and judgments we have about the data is a learned skill. Here are a few examples to help.

When you hear the evaluation	Look for the underlying data
The build is constantly broken	The build was red on 7 occasions
Assets are ugly	I think customers won't like them
The team is way too big	Three people were added to the team

The words above that are clues that we're in evaluation mode are "constantly", "ugly", and "way too". All of those words have many meanings, and those meanings are not the same for everybody.

4. Examine the data, and think it through

Once we gather up enough data, and make it visible to the team, we can start to look for patterns together. "Oh! 19 of the 27 stickies are about the new design process!" Perhaps the team will, at this point, begin to classify the bits of data they're looking at as helpful or unhelpful. Now's the time for those feelings to be examined along with the data.

Many of the retrospectives I see in the wild skip the earlier stages, and start here. The team gets together, without deciding what data they will examine, and begins to write down their evaluations of whatever topics come to mind, and stick them on the wall. Judgment is embedded in this method, as each post-it goes into a plus or minus column.

Deciding what to change before you've considered the data and thought it through as a team bypasses analysis, preventing us from coming up with surprising new information, and solutions that go beneath the surface. We stay at symptom relief.

5. Decide on actions to take

Once the team has thought through the data that's been addressed, noticed patterns, and considered how each factor affects the team, it's time to decide what actions to take.

One way to do this is to form a hypothesis, and then go about testing it. We can choose one thing to change, think about the expected outcome, and then later look at the results. When we come back to revisit the hypothesis, we ask "Did we make the change we intended?" (maybe "What prevented it?") and then "What was the effect?"

6. Close

Retrospectives can raise topics that carry emotional weight. Rather than letting the discussion fizzle, I like to plan a closing, to help us shift our attention to other work. Sometimes, concerns are still present. Someone will have had trouble choosing just one action to take, or things that we haven't found time to talk about can continue nagging at a teammate's attention. They may fear their concerns will never be addressed.

In those cases, I sometimes close by asking folks to name how they're feeling and any lingering concern. It can be helpful to find a place to capture those concerns. This lets us address them later, and it can also help people find closure for the retro, rather than leaving with frustration. It's easier to transition to regular work at the end of our retro if I can ensure everyone feels heard.

With a new team, or a team without much skill in communication, this sort of retrospective can be a little disorienting. After the first several retros, when the team begins to feel that they'll be heard — that they'll hear *each other* — and that the team will come up with useful actions to try and experience real improvement, trust increases, and retrospectives become more and more effective.

conflict

logic games

There is a logic game I enjoy. It involves numbers, and pictures, and figuring out which bits of a grid get marked and which don't. It's not difficult. Each one is a simple puzzle that passes the time, sort of like an easy sudoku.

I discovered something recently that really surprised me. Having been upset about something, I'd picked up my little game to unwind. And when I got to the end, there was a contradiction. My grid was a failure, and I didn't know why. I'd done all the logical steps, but at the end, things didn't line up.

I didn't think much about it until it happened several times in a row. Puzzled, I set it aside.

But the next time I tried to play it while I was sad (maybe depressed) or full of anxiety, I had the same experience. And I wondered why. Nothing like a good mystery to distract from sorrows!

I came to realize was that it was a real block. I am apparently *not rational* when I'm experiencing strong emotion.

Woah.

This, I realized, *is gonna have a lot of implications for how I interact with other people.*

Of course, nobody's completely rational. All sorts of thinking biases have been discovered. But I do like to think I can work out logic!

This experience told me that I really do need to back off and take care of myself, rather than going in fighting. I need to get to a place where I am pretty serene, if I want to have a hope of approaching the world with the sort of calm awareness I hope for.

And it suggests that other people may need tender care, as well, if we're going to make good decisions when feelings are strong!

avoiding conflict

It's not universal, but I've noticed on many of my teams that the folks who seem most opposed to paying attention to feelings at work are the folks with some of the strongest feelings. It's not surprising — strong feelings are scary. If I know I sometimes feel like throwing things, I might think that avoiding any attention to feelings will prevent that sort of thing from happening. If seeing someone else cry scares me, I'm going to do my best to prevent them from experiencing (or expressing) sadness.

But feelings affect us, whether we're talking about them or ignoring them. My trouble with the logic game illustrates that. Since that discovery, I've noticed other times when when strong emotion correlated with my not understanding something that seems like it shouldn't be hard.

Avoiding conflict can be tempting as a way to take care of folks and keep people happy. The problem is that conflict is still present. We can't prevent it, we can only prevent *dealing* with it, and moving forward together.

vulnerability & risk

In 2010, Brene Brown presented a TED talk about vulnerability that got people very excited. She told us that she'd done studies about what resilient — in her words "whole hearted" — people have in common. What she found was a willingness to be open, to be vulnerable. Her "whole hearted" people share a willingness to expose their uncertainty, to be less-than-perfect, and as a result they are able to connect with other people.

This was very much in line with what I was thinking and talking with colleagues about in 2010. One thing that kept coming up in conversations is how sometimes the lovely descriptions of the whole-hearted life leave out the fact that "vulnerable" actually means "able to be hurt". There's risk of conflict, discomfort, exposure.

The empathetic team has decided that those risks are worth the rewards of connection, effective decision-making, better productivity, and enjoyable work.

connection

Remember the hypothesis I suggested back in the chapter about empathy, and again in retrospectives?

What if every person you come across is doing the best they can to meet needs you can understand and recognize in yourself, even when their strategies are not to your liking?

I invited you to try it on. To see how it works when you apply it as a model for making sense of the world. How has it worked? What have you discovered?

When I feel strongly about things, it's hard to see things this way. And yet, I haven't found any model I like better.

One interesting effect that comes from living with this needs-based model of human behavior is that in any disagreement, it becomes much more difficult to see one person as simply right, and the other as simply wrong.

Another thing the needs-based model tells us is that it's possible, if we open up to each other and listen well, we can understand where someone else is coming from. When anger gets replaced by understanding, conflicts are resolvable.

team decisions

For a technical team, a constant stream of decisions needs to be made. Often, they're pretty easy. What library should we use? What sort of API will work best? Emotions might be elevated when we're deciding whether to test drive code (or what test driving actually means), whether we're refactoring enough or too much, or how long a method is too long.

When writing tests, how much do we mock? Should we change from Java to Clojure? Or Haskell?

All you're learning about listening, empathy and facilitation will help your team to make these decisions. Keeping team communication open, and being willing to talk things through goes a long way.

At some point, though, your team may face a personal conflict that gets in the way of your ability to deliver. As someone the team relies on for support and leadership, you may find yourself in a pretty awkward place, dealing with a lot of difficult emotions.

On one team I worked with, things got so bad between two team-members that they wouldn't pair, wouldn't talk to one another. Finally, things came to a head during a morning standup. They were openly arguing, interrupting each other, and blaming.

We started on a path to recovery using *listening method* for conflict resolution. Maybe it will come in handy for you sometime as well.

the listening method

The listening method has a few ground rules.

1. Both people must want to try it, in order to understand and be understood.

2. Both agree that when they're talking, they'll talk only about their own experience.

3. They agree that when it's their turn to listen, they'll only speak to ask questions to clarify what they're hearing.

4. All feelings are welcome. And each person listens for only as long as they're willing. This means it's okay to get angry, and it's also okay for the other person to want to end the session.

When the two people agree that they'd like to try the listening method to resolve the conflict, the two of them, with a facilitator if they wish, sit face to face. Then they choose who will go first.

The person who speaks first says "One thing I'd like you to know is…" and completes the sentence. It's often helpful to keep it brief, knowing there will be more opportunities to add one small thing at a time.

Second listens. Then they say "Thank you." And after this, they respond by *reflecting back what they've heard*, and they *ask whether they've understood correctly*. "Did I get that right?"

And then "Is there more you'd like me to know?"

This continues until the first person is ready to say "not right now." Then, it becomes the second person's turn to share.

This repeats until there a shift happens. The two have a sense of understanding one another. There's connection, rather than only opposition. For example, you may see signs that the two are beginning to explore ways for both of them to be happy.

I've found the difficult part here — when things are heated — is the temptation of the listener to defend, or to argue. The solution to that is being heard. Since only one person can speak at a time, at least one of the participants needs enough empathy (or self-empathy) in advance to let them listen, even when it's hard.

self-care, team care

Finding kindness, patience, and curiosity about the other person's experience is much easier when you're in a good place. You can get there by taking care of yourself — talking things out with a friend, taking a walk while your head clears, or even getting a massage over lunch. And you can support your team in doing those same things to care for themselves, so they're ready to face conflict with care. You may even find that there's a form of self-care that applies to the whole team. Notice what your team needs in order to be able to show up with courage, when the time comes.

listening method quick reference

Ground Rules

1. Both want to use this method to understand and be understood.
2. Both agree that when they're talking, they'll talk only about their own experience.
3. They agree that when it's their turn to listen, they'll only speak to ask questions to clarify what they're hearing.
4. All feelings are welcome. And each person listens for only as long as they're willing. This means it's okay to get angry, and it's also okay for the other person to want to end the session.

A facilitator is optional. The participants sit face to face, and choose one to speak first.

Speaker: What I want you to know is…

Listener: "Thank you" or "Could you clarify…" or "I think you're saying… Is that right?" Listener: Is there more?

Speaker: Not right now.

Switch, and repeat until there's a shift in connection.

mentoring

making mistakes

It's 6pm on Tuesday. You've spent the last day and a half trying to track down a bug. And just now, you discovered that the problem was you. It turns out that you'd left another copy of the system running in the background, which was messing with your tests.

Yesterday at standup, you confidently said you'd have the bug figured out in a few minutes and get back to your task. But it took the whole day. You screwed up. What do you do?

What do you say at standup?

How do the words you choose affect you? And what effects do your words have on your team?

When I'm working with a team, as an empathetic leader I notice a lot of different responses to emotions that come up around mistakes. On some teams, I see mistakes discussed in hushed tones, eyes down. Body language suggests to me a feeling of fear and constriction.

I want my team to learn together easily, so I hope to encourage a relaxed approach, which becomes visible to me in the form of laughter and gentleness. People ask questions readily, openly discuss ideas (that might be wrong) and resolve problems together as they come up.

When there's a mistake, or someone is telling themselves they screwed up, noticing what happens then can tell me a lot about what might be standing between me and the productive and fun team I want. So I observe closely. And I model the kind of culture I want to work in.

When I screw up, I first want to be sure I'm open and real about it. That means I acknowledge the facts, without embellishing them with drama. I say "I broke the build", and maybe add that I'm a doofus. What I don't add is a lot of sorrowful self-loathing commentary. My intention is to let us all make what mistakes we make, and recover with agility and resilience.

If I do feel horribly self-critical, sometimes I'll share that with the team, but only if we've grown to a place where we can put real attention to it. I'll want them to know that those feelings are coming up, and that *I don't buy into them*. That my world view doesn't actually support me being useless, or stupid, or whatever I'm telling myself. That my brain lies to me sometimes.

If, on the other hand, the team doesn't have a lot of history yet, and we're beginners at being a team, I'll work though my feelings of horribleness with a friend, outside work. I wouldn't want to accidentally convince my team of the lies my brain is telling me. If I convey that I'm terrible because of a mistake, I'm not just modeling it, I'm promoting the theory that a mistake makes a person terrible. If folks accept my theory, then they have to be terrible for making mistakes as well.

fast feedback

Perfection is not a thing humans can do. As we work to solve complex problems in a complex context, we must try various things, and notice the results, in order to get closer to where we want to be. Because we know we'll take "wrong" turns, try things that don't work, we get good at taking small steps and getting fast feedback.

One of the ways we do this is by delivering a tiny bit of functionality at a time, creating a finished product every week or two. Another way is through test-driven development

(TDD).

When I'm pairing with another dev, I can model being a doofus, just like in standup or meetings (whether ad-hoc, at the whiteboard, or scheduled, like planning). But I also model curiosity, comfort with uncertainty, and making our code commits tiny, so I'm happy to revert whenever things aren't working out.

When practicing TDD, we write the very smallest test we can think of, and the smallest bit of code to make it pass. Then we refactor to keep things DRY[6] and easy to understand. We do this in small enough steps, committing each time, so that if we go for twenty minutes without a commit, that's a good clue that we're headed in the wrong direction.

If I'm on a team where the word "wrong" has lost its power to scare us, we can simply say, "oh, this isn't working", reset, and start that bit over. All while having fun!

pairing

Pair programming is one of my favorite ways to mentor. We learn from one another, and learn to get better at the things

---

[6]Don't Repeat Yourself, shorthand for the idea that duplication in software is trouble.

we're good at, and less bad at the things we're bad at.

Pairing can be uncomfortable. Until you get used to it, making all your decisions out loud, with someone sitting right there, can be intimidating. An empathetic leader can help a lot by helping their pair have a fun pairing experience. The folks who tell me they like pairing and want to pair are the folks who have had at least one really good pairing experience. Accomplish that one success, and you're a long way toward loving it, and helping your coworker love it.

do the wrong thing

When you find yourself pairing with someone, and you know the right way to go, do you ever find they want to do something entirely different? Sometimes, when I know I'm right, I allow the desperate urge to win to go unheeded, and try my pair's approach. Pretty often, one of two things will happen. Either we will discover my pair's idea was awesome, in which case, by resisting the temptation to cling to my idea, I've avoided looking foolish. Or, occasionally my pair's idea is as faulty as I thought it was, and it becomes apparent, without my having been stubborn and obnoxious about it.

sit on your hands

Pairing with someone who knows less than you do about the tool set, the tech stack, or the language will probably slow you down. That can be a lot less frustrating if you take seriously these three ideas.

First, that bringing them up to speed is your best way to help your team be more productive quickly.

Second, that slowing down is quite possibly a good thing.

Third, that whatever you know more about than they do, they have some things to teach you, as well.

But I'll say it again: don't take my word for it. Try it. Watch and see what happens. What happens when you explain a bit of code to them? Do you discover that your names are lovely and your code understandable? Or do you find it gets much better when you have someone at your side? Do you know all there is to know about your IDE[7]? Every time I pair in Vim, we each learn a new trick.

When you're with someone who's learning, it can be tempting to just start typing, and hope they can keep up. But if you do that, you're missing out on an opportunity.

---

[7]Integrated Development Environment. Editor and related tools for coding.

My friend Geepaw Hill, who's also a software coach, advises coaches to "sit on your hands". This means letting the noobie "drive" at the keyboard, and as the "coach", you take the job of keeping the conversation going, modeling great pairing, and exploring questions that come up.

It's something to try.

self-care

Sometimes it gets hard. You wake up late, have a flat tire, you know the kind of day. And then you're pairing and you're just gritting your teeth. What do you do?

Think of this book as a fractal. The whole thing can be applied on the team level, but you can also zoom in to the pair level (and probably out to the organization level, to some extent). The chapters about listening and conflict could be helpful.

Or, try this: When you feel yourself thinking "wtf?!" or "why would somebody do that?!" see if you can convert the "why" part into a real question in your mind. If you can convert your grumpiness to curiosity, you might find things shift to a much more comfortable place.

Of course, that can be hard, especially on a rough day.

When I want to show up as a caring coach, an empathetic leader, self-care is vital. If I'm stressed, I'm not able to show up the way I want to, to listen, and to be curious. It all breaks down. So I care for myself, allow myself to have needs, and to get them met.

How do you respond when you're under stress? Do you know the signs that tell you you're running out of patience? Getting coffee, sitting quietly, taking a walk (or a run), or working on some fun code are some ways leaders take care of themselves. What helps you recover? If you don't know, this might be a good time to begin to figure it out.

journey

In the first part of this book, we covered inner work that is the grounding for empathy. It's hard work, easier to talk about than to practice.

The skills in part one are skills you'll come back to again and again, as an empathetic leader. And if you're anything like me, they'll never be something you can take for granted. My skills suffer whenever I'm not well-supported. Maybe I'm dehydrated, or I haven't slept well. Or maybe something's come up that touches on my most tender places. Someone has angrily questioned a decision I was already anxious about. Or I've just been rejected for a project I applied for, and someone makes a comment, and I snap.

self-care

When I'm feeling threatened, or weakened in any of these ways, I become less empathetic, less mindful, less able to listen. To get back to being more like I want to be, there are a few things I've found helpful. Maybe they'll help you, too.

self-empathy

What happens when you find yourself angry? Irritated at someone on your team, feeling hopeless about solving problems that come up?

Do you also get frustrated *about* that frustration? About having let yourself down?

Here's a new twist on an idea you've become familiar with:

> *What if you are doing the best you can*
> *to meet human needs we can all understand*
> *and recognize in the people we love, even when*
> *your strategies are not to your liking?*

What would it mean to forgive yourself? What would it mean to be your own best friend?

befriending

There is a practice I call "befriending". It starts with getting comfortable. Maybe going through the "noticing" exercise, back in the chapter on mindfulness. The practice is to sit *with myself*, listening to what's happening inside me. I pay the kind of attention that I pay to a close friend. I listen to me, with care and empathy. I practice being a friend to me.

joy

Sometimes you may hear advice about "gratitude practice". When I first heard this, I guess I figured it was some magical esoteric practice. But it's not. Let me break it down for you.

The habit of noticing what is pleasing to you changes your thoughts, your expectations, and your creativity.

That's not so magic, is it?

In just about any moment, I can find things I don't like and things I do like. When my attention is on the things I don't like, sometimes my abilities suffer — to think clearly, to question the stories in my head, to creatively look for solutions. I can find myself more stubborn, more angry, and less effective.

If instead I put some attention to noticing where the bits of joy are, things shift.

For some folks, this practice is gratitude. For other folks, gratitude doesn't make sense. It might help to think of a *joy journal*. Find some time each day, or when you feel you could use it, to write down five things you're happy about, right. Five joys.

simple self-care

What do you like to do to wind down? Play video games?
Take bubble baths? Hike in the woods?

Sometimes, I have put off doing the things I love because
I "have to" work, to be responsible. But a little kindness —
well, you know what it can do. Will you show yourself the
same kindness you hope to show others?

At work, what if you decide it's okay to take fifteen minutes
to go outside and breathe? Listen to one song in your
headphones? Or get coffee and just sit for a minute.

A little kindness. For you, too.

next steps

It's my hope that this Little Guide has given you some ideas to think about and things to try on your way to becoming the empathetic technical leader you want to be. What comes next?

Here are some ideas to consider:

- Join a communication skills practice group (or create one).
- Try the listening method with friends, away from work.
- Form a practice group with other leaders in your company.
- Start a book group for empathetic leadership.
- I'd love to hear about your experiences, at etl@maitria.com.